C0-DYB-061

SAVING ANIMALS

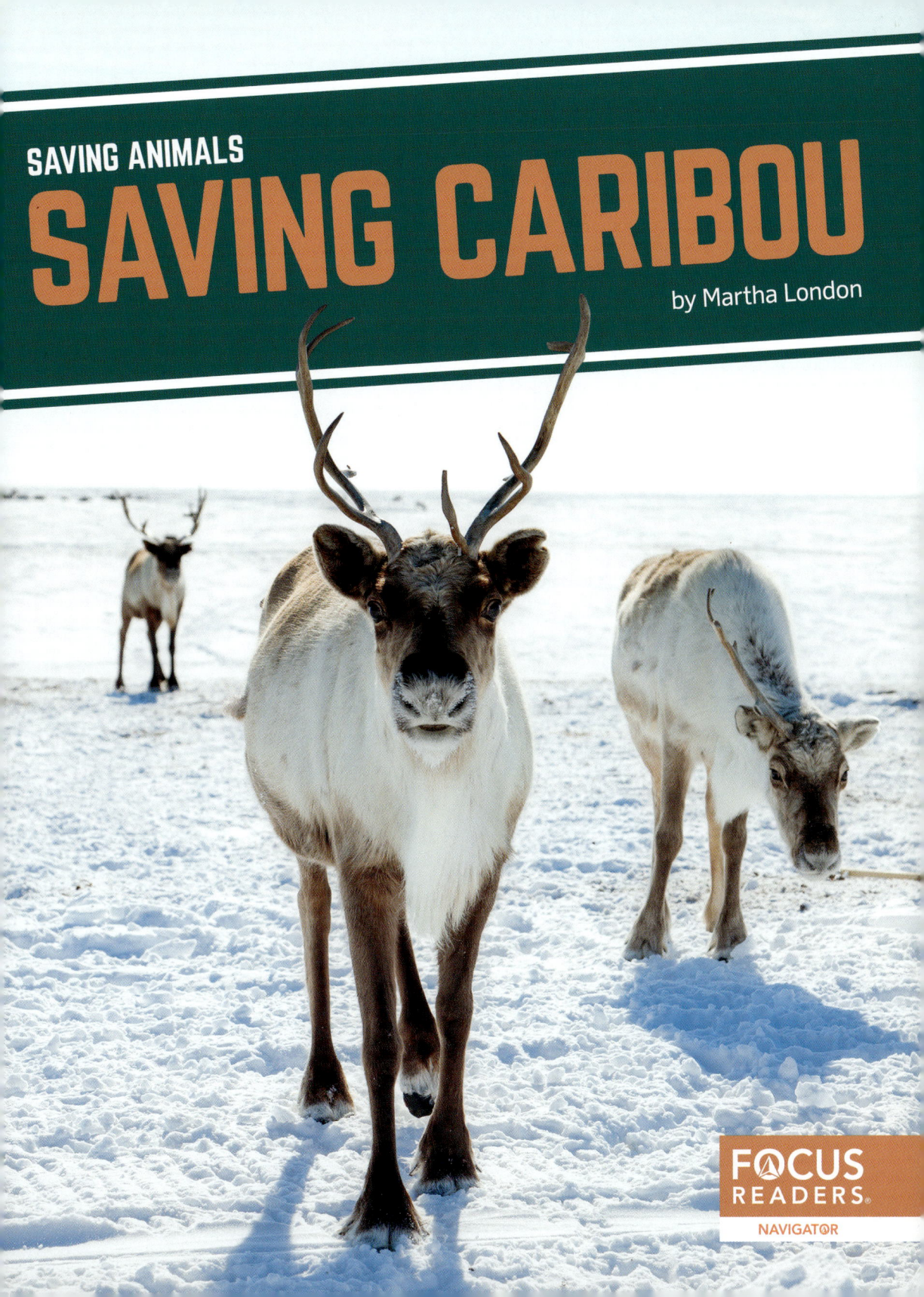

SAVING CARIBOU

by Martha London

WWW.FOCUSREADERS.COM

Copyright © 2021 by Focus Readers®, Lake Elmo, MN 55042. All rights reserved. No part of this book may be reproduced or utilized in any form or by any means without written permission from the publisher.

Focus Readers is distributed by North Star Editions:
sales@northstareditions.com | 888-417-0195

Produced for Focus Readers by Red Line Editorial.

Content Consultant: Karen H. Mager, PhD, Assistant Professor of Environmental Sustainability and Biology, Earlham College

Photographs ©: Shutterstock Images, cover, 1, 4–5, 7, 10–11, 15, 21; iStockphoto, 9, 13, 27; Montana Fish, Wildlife and Parks/AP Images, 16–17; Podgorny Igor/Zuma Press/Newscom, 19; Bob Wick/Bureau of Land Management, 22–23; Red Line Editorial, 24; Ian Langsdon/EPA/Rex Features, 29

Library of Congress Cataloging-in-Publication Data
Names: London, Martha, author.
Title: Saving caribou / Martha London.
Description: Lake Elmo, MN : Focus Readers, [2021] | Series: Saving animals | Includes index. | Audience: Grades 4-6
Identifiers: LCCN 2019057428 (print) | LCCN 2019057429 (ebook) | ISBN 9781644933855 (hardcover) | ISBN 9781644934616 (paperback) | ISBN 9781644936139 (pdf) | ISBN 9781644935378 (ebook)
Subjects: LCSH: Caribou--Conservation--Juvenile literature.
Classification: LCC QL737.U55 L595 2021 (print) | LCC QL737.U55 (ebook) | DDC 599.65/8--dc23
LC record available at https://lccn.loc.gov/2019057428
LC ebook record available at https://lccn.loc.gov/2019057429

Printed in the United States of America
Mankato, MN
012021

ABOUT THE AUTHOR
Martha London writes books for young readers. When she isn't writing, you can find her hiking in the woods.

TABLE OF CONTENTS

CHAPTER 1
In the Wild 5

THAT'S AMAZING!
Antlers for All 8

CHAPTER 2
Habitat Helpers 11

CHAPTER 3
Threats to Caribou 17

CHAPTER 4
Protecting Caribou 23

Focus on Saving Caribou • 30
Glossary • 31
To Learn More • 32
Index • 32

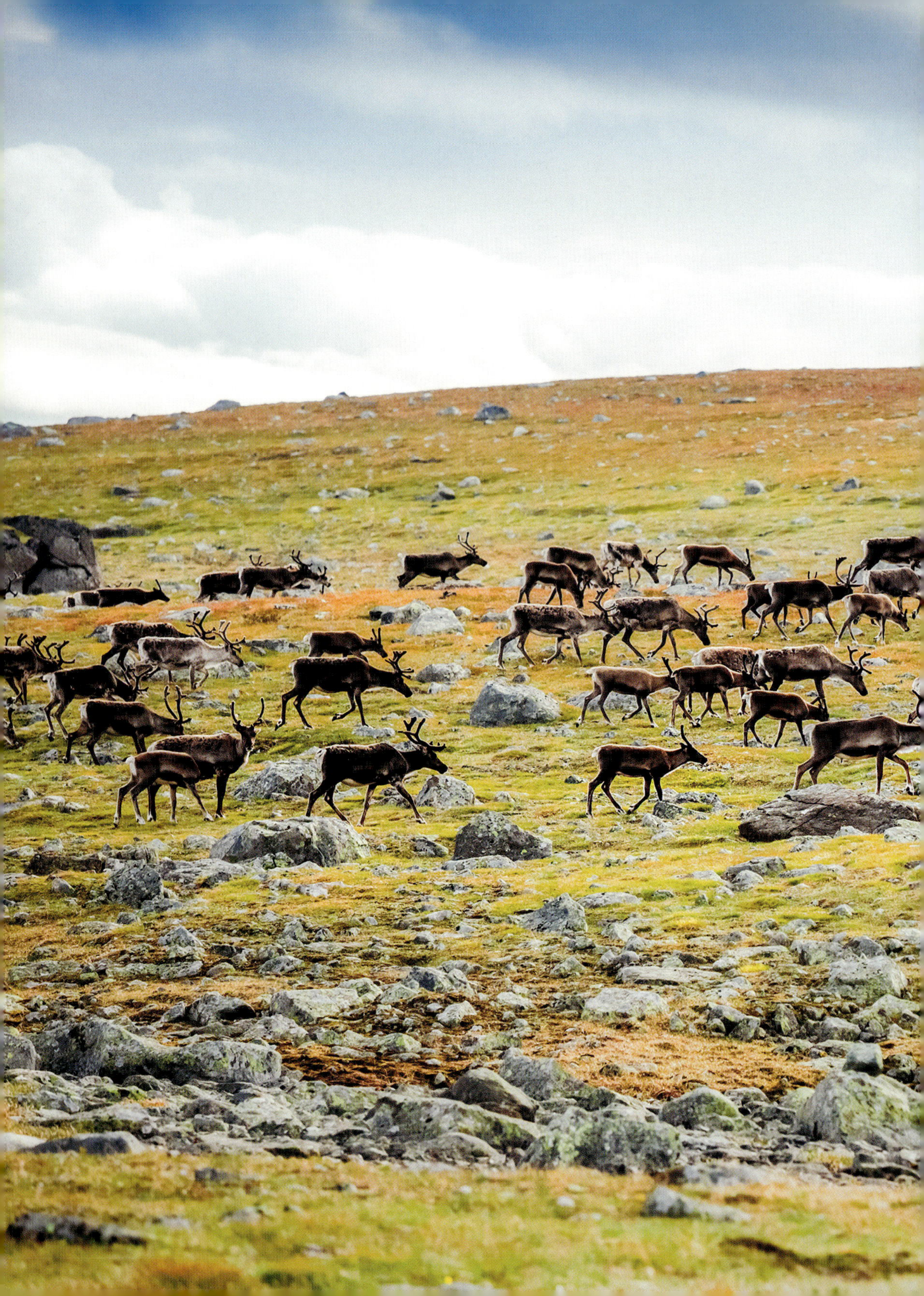

CHAPTER 1

IN THE WILD

Caribou are a type of deer. They can be found in North America, Europe, and Asia. Some caribou live in the Arctic **tundra**. Others live in **boreal** forests of the subarctic. This region is south of the Arctic. Some deer **migrate**. Others do not.

Caribou live in cold **habitats**. Their bodies are suited to live in harsh climates.

Caribou have lived on Earth for more than 1.5 million years.

For example, caribou's fur coats block water and strong winds. As a result, caribou stay warm even during blizzards.

Caribou do not eat meat. They mainly eat lichen, a plant-like organism. During winter, snow <u>covers</u> the ground. Even so, caribou are able to find food. The deer

CARIBOU MIGRATION

The Porcupine caribou herd has one of the longest migrations of any land animal on Earth. This herd may travel more than 1,200 miles (2,000 km) each year. During the warmer seasons, the herd travels north into the tundra. When winter arrives, the herd moves south into the boreal forest. Snow is softer there. It is easier to dig for food.

have wide, sharp hooves. They use their hooves and antlers to dig through the snow. They eat the plants hidden under the snow. Caribou are always on the move. So, they do not eat too many plants in one place. In this way, these animals help keep their habitats healthy.

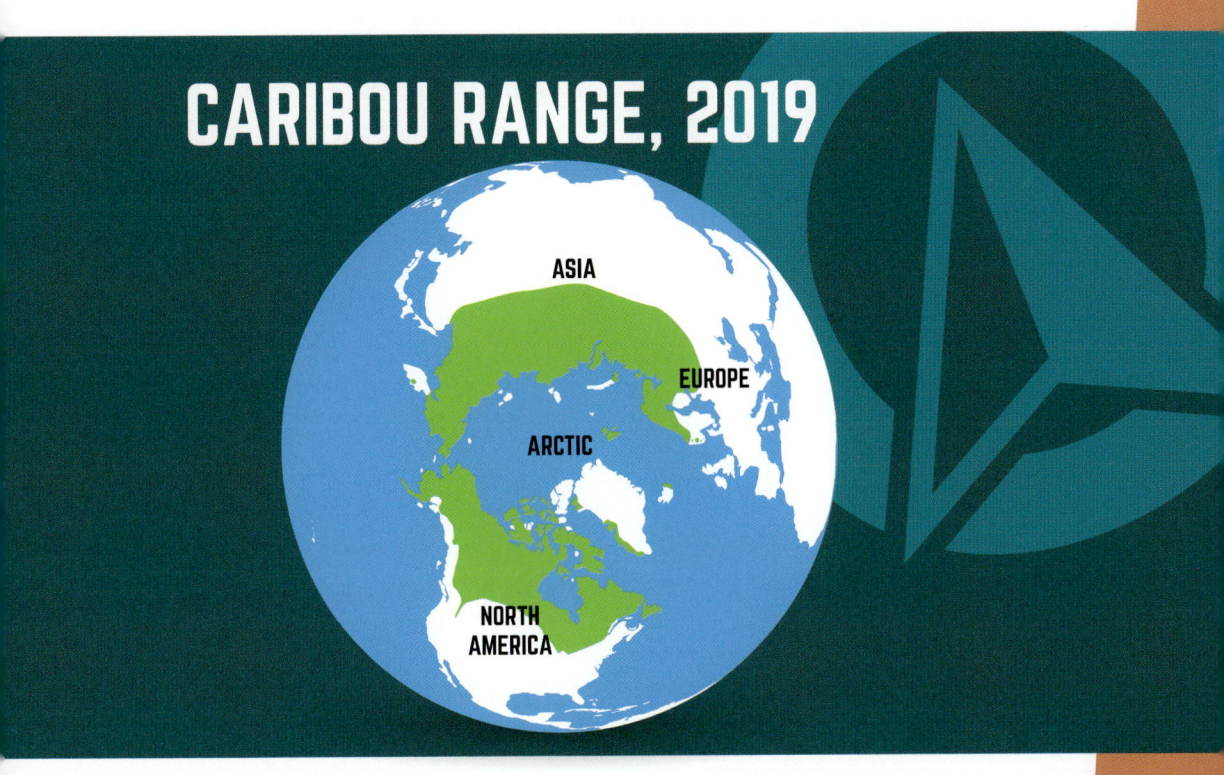

CARIBOU RANGE, 2019

ASIA

EUROPE

ARCTIC

NORTH AMERICA

THAT'S AMAZING!

ANTLERS FOR ALL

For most kinds of deer, only male deer grow antlers. Caribou are different. Both male and female caribou grow antlers. Like other deer, caribou shed and regrow their antlers every year. But male and female caribou shed at different times.

Male caribou begin growing their antlers in March. Their antlers stop growing in September. They fall off in November. The antlers of female caribou start growing in June. Female caribou keep their antlers over the winter.

Scientists believe that female caribou may use their antlers to protect their food. Female caribou tend to become pregnant during the winter. Pregnant caribou need to eat a much larger amount of food than other caribou. Thanks to their antlers, they can keep other caribou away.

Female caribou tend to have smaller antlers than male caribou.

As a result, pregnant caribou get enough to eat. Female caribou give birth in the spring. Shortly after, they shed their antlers.

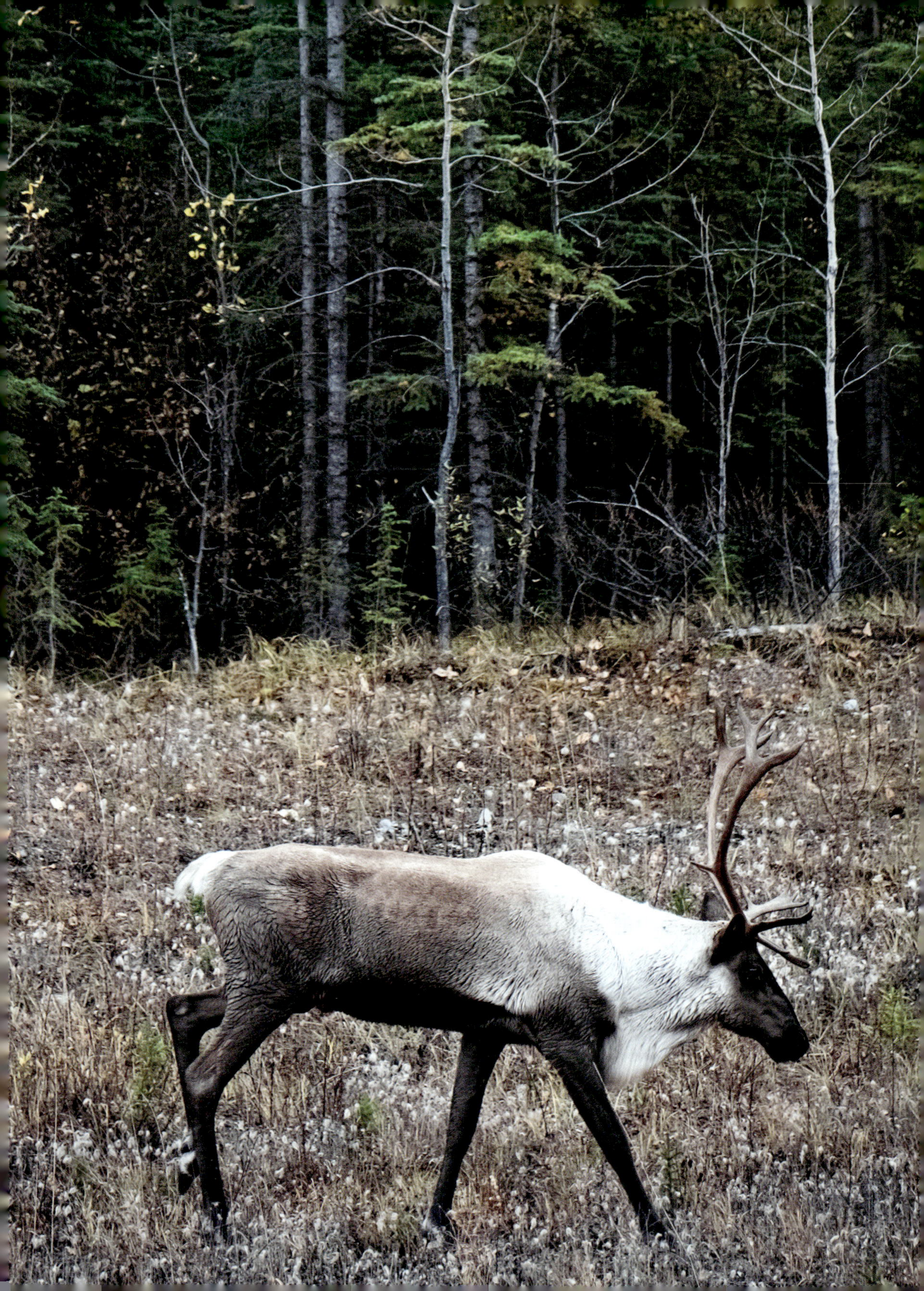

CHAPTER 2

HABITAT HELPERS

Caribou live in two types of environments. They act differently based on where they live. Forest caribou tend not to migrate. But tundra caribou do migrate. The tundra has few trees. During winter, tundra caribou move closer to boreal forests. These forests provide the deer with shelter.

A caribou walks through a boreal forest in British Columbia, Canada.

Without caribou, tundra and boreal **ecosystems** would change. For example, caribou spread their droppings as they travel. These droppings include plant seeds and nutrients. Seeds and nutrients help new plants grow. Caribou droppings benefit both tundra and forest areas.

In addition, other animals rely on caribou. For instance, Arctic foxes live in the tundra. They follow caribou during the winter months. Caribou use their hooves to uncover plants beneath the snow. Mice hide under the snow. Other rodents do, too. When caribou dig, these animals have to move. As a result, foxes find food much more easily.

Arctic foxes can live in areas with temperatures as low as −58 degrees Fahrenheit (−50°C).

Caribou are also food for certain animals. Wolves and bears eat caribou in forests. Without caribou, wolves and bears would move to other areas to find new prey. With these predators gone, other animals would take over the forest.

This change could make the ecosystem unbalanced.

Ecosystems need balance to be healthy. All types of plants and animals help keep this balance. If there are too many predators, prey disappear. If prey

CARIBOU AND NUNAMIUT PEOPLE

Certain **Indigenous** peoples have depended on caribou for thousands of years. For example, the Nunamiut people are a part of the Inuit peoples. The Nunamiut live in northern Alaska. They hunt caribou. They use all parts of the caribou. For example, they eat the meat and make clothes with the furs. For some villages, caribou is the only source of meat. They consider caribou to be sacred.

Wolves tend to hunt caribou in packs.

disappear, the plants they eat may grow too much. The spread of those plants can stop other plants from growing. Over time, the area may no longer be able to support certain animals.

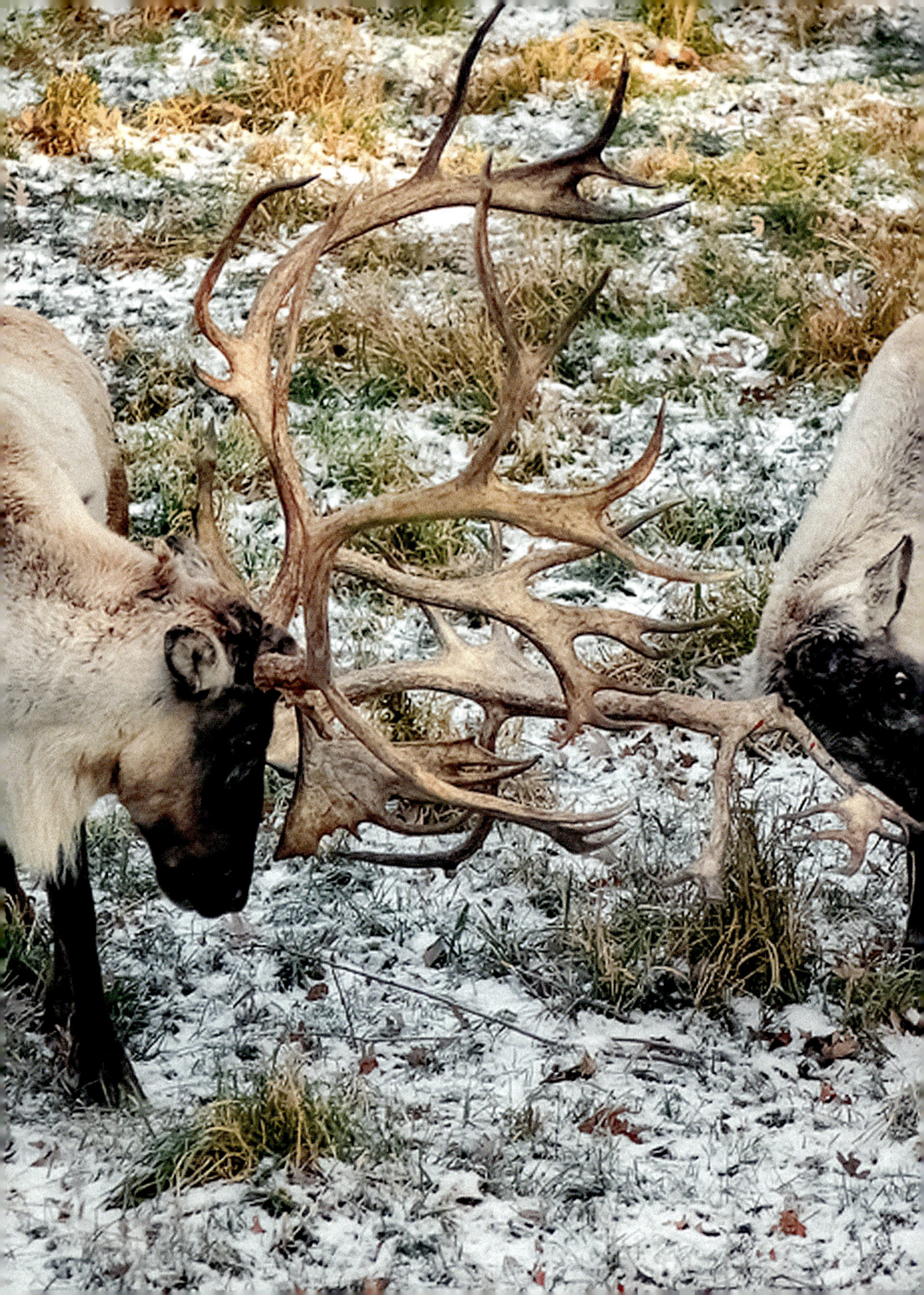

CHAPTER 3

THREATS TO CARIBOU

Caribou face many threats. Caribou populations have been dropping for years. Their numbers do tend to rise and fall. But scientists say the current drop is different. Some herds have already died out. For example, caribou herds used to live in Idaho and Washington. But they have disappeared from these areas.

People used to see caribou in Montana. In 2019, scientists moved the last of the herd to Canada.

One major threat to caribou is habitat loss. The main causes of habitat loss are logging and mining. These industries are important in Canada, Russia, and other places. Sometimes companies clear forests to make room for new houses. Other times, companies chop down trees to use the lumber.

In addition, companies mine and drill different materials. These materials include coal, oil, gas, and gold. Workers dig into the ground. They create huge pits in the land. After a mine is closed, it takes years for the land to recover. In that time, caribou move far away. They do not live in areas where the land is disturbed.

A mining company clears a forest in northwest Russia, where a threatened caribou herd lives.

Logging and mining do more than destroy habitats. They also divide habitats into smaller sections. This process is called fragmentation. Building roads can have the same effect.

Fragmentation hurts caribou. That's because caribou need large amounts of space to avoid predators. In the tundra, caribou are able to run for miles at a time.

And in forests, trees protect the herd. But caribou tend to avoid roads. As a result, they cannot avoid predators as easily.

Habitat loss causes other problems for caribou. When forests are cleared, new kinds of plants grow. These plants attract other animals, such as moose and deer. The new animals compete with caribou for food. They also attract more wolves.

One of the largest threats to caribou is **climate change**. This crisis harms caribou in many ways. For instance, temperatures in caribou habitats have warmed by 4 degrees Fahrenheit (2.2°C) since 1890. This extra heat brings more insects. Caribou will run for miles to escape

A single swarm of Arctic mosquitoes can kill a baby caribou.

mosquitoes. When caribou run, they eat less food. Then they are not as healthy when winter comes.

Also, grazing and calving grounds are blooming earlier than before. But caribou are still arriving at these areas at the same time each year. So, they arrive late. They have less food to eat. As climate change gets worse, caribou will face greater threats.

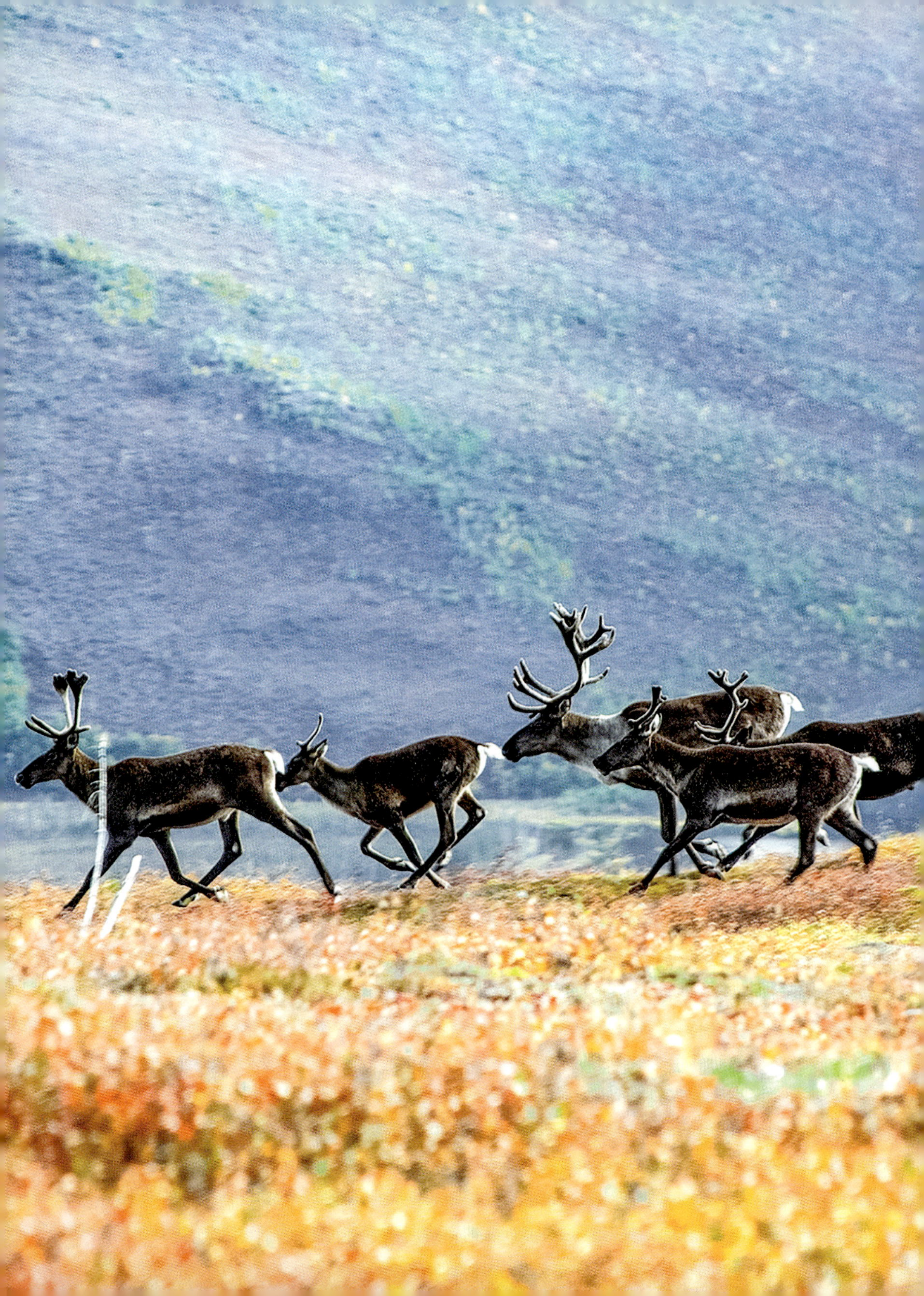

CHAPTER 4

PROTECTING CARIBOU

Between 1995 and 2018, the number of caribou dropped by more than half. Five North American herds will not grow back. Still, many groups are trying to help. Some efforts are working. For example, the Fortymile herd has grown.

In the 1990s, the governments of the Yukon territory and Alaska took action.

The Fortymile herd of caribou can often be found in Alaska's Steese National Conservation Area.

Both governments limited hunting this herd. In Alaska, hunters agreed to hunt only 150 Fortymile caribou per year. In the Yukon, hunters agreed to stop hunting the herd completely. By the 2010s, the herd's numbers were much higher than in

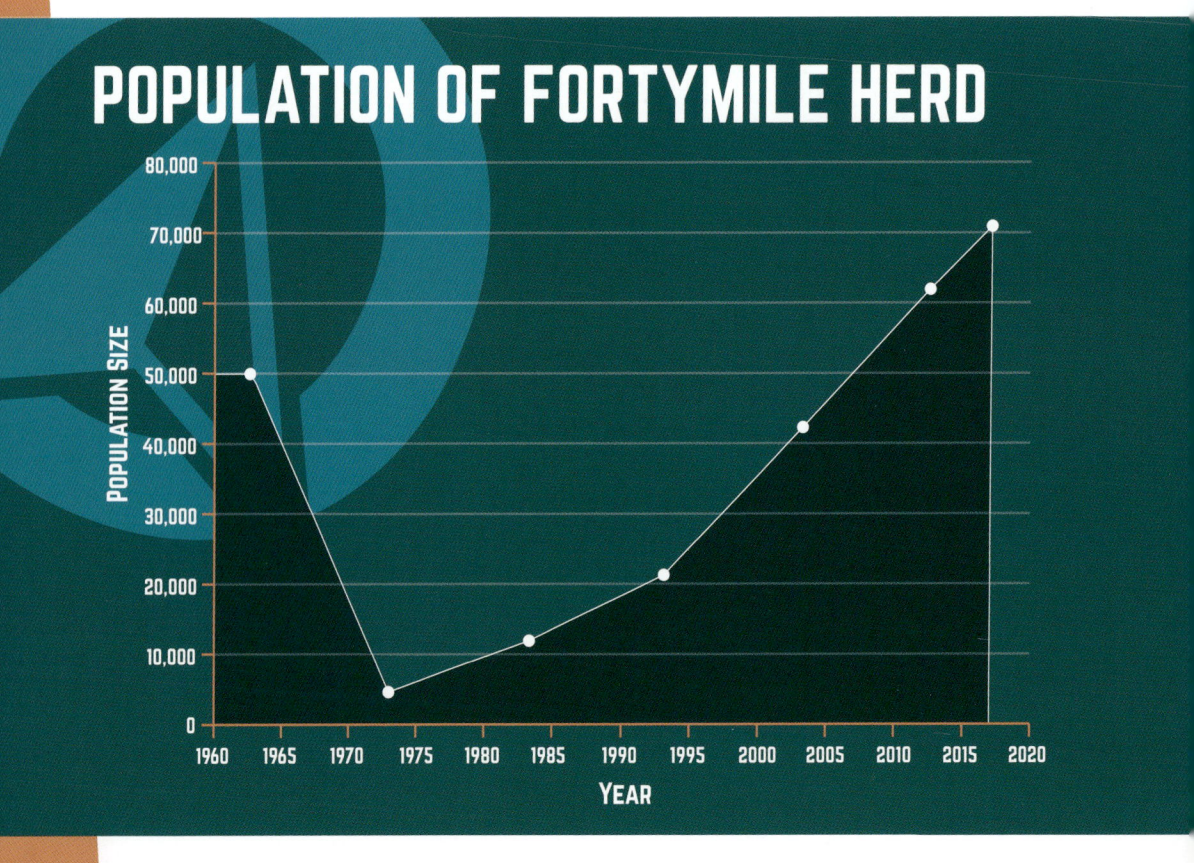

POPULATION OF FORTYMILE HERD

the 1970s. Scientists believe the limits on hunting may have been one reason why.

Scientists are studying caribou herds. To do so, they catch members of a herd. First, they fire nets from helicopters. Next, they place collars on the captured caribou. Finally, they release the caribou. Scientists can track the collars. In this way, they learn more about where the herd goes during the year. They also learn where the herd gives birth.

This information helps scientists. They learn which areas are most important to protect. Without those areas, fewer baby caribou would survive. The herd's numbers would not be able to grow.

Other groups are trying to stop habitat loss. They want more protected areas for caribou. In protected areas, people are not allowed to do anything that harms the environment. For example, companies cannot chop down trees in these areas. Forests can keep sheltering caribou.

KASKA DENA

Kaska Dena are a First Nations people in northwestern Canada. They live in an area that is home to several caribou herds. In June 2019, they put forward a plan. The plan would protect much of the area. Canada's national government supported this plan. In September 2019, Kaska Dena leaders met with members of the local government. They needed that government to approve the plan.

Gaspesian Provincial Park in Quebec, Canada, has protected an endangered caribou herd since 1937.

Governments can protect areas better than **conservation** groups. For example, governments can make harmful activities illegal. But they sometimes act slowly. One example is Alberta, Canada. The government said it would protect caribou homes. It said plans would be ready by 2017. But in 2019, Alberta's plans still were not ready. Conservation groups looked to Canada's national government. They hoped it would force Alberta to act.

Many Indigenous groups are also helping caribou. For instance, the Déline people came up with a plan in 2017. These people often depend on caribou for food. Their hunting had not caused caribou numbers to drop. But hunting fewer caribou may help herds grow.

For this reason, the Déline nation's plan limited caribou hunting. People would get food from other animals. And communities would share more with one another. These communities would also keep track of the herd's health. The government of Canada's Northwest Territories agreed with the Déline's plan. It used the plan for the entire territory.

A Déline chief leads a protest outside a 2015 conference on climate change.

However, none of these actions will slow climate change. Nations must make huge changes. They need to cut how much oil and coal they use. They must use other kinds of energy, such as wind power and solar power. As of 2020, few nations were changing fast enough. People need to demand these changes from their governments. Caribou face a variety of dangers. Even so, many people are working hard to protect them.

FOCUS ON
SAVING CARIBOU

Write your answers on a separate piece of paper.

1. Write a letter to a friend that explains the key ideas of Chapter 3.

2. Do you think current efforts are enough to save caribou? Why or why not?

3. What do caribou mostly eat?
   - **A.** wolves
   - **B.** lichen
   - **C.** moose

4. Why might governments be best at creating protected areas for caribou?
   - **A.** Governments are the most motivated to protect caribou.
   - **B.** Governments know the most about how to best protect caribou.
   - **C.** Laws can be the best way to make sure people follow certain actions.

Answer key on page 32.

GLOSSARY

boreal
Having to do with forests in northern climates.

climate change
A human-caused global crisis involving long-term changes in Earth's temperature and weather patterns.

conservation
The careful protection of plants, animals, and natural resources so they are not lost or wasted.

ecosystems
Communities of living things and how they interact with their surrounding environments.

habitats
The types of places where plants or animals normally grow or live.

Indigenous
Native to a region, or belonging to ancestors who lived in a region before colonists arrived.

migrate
To move from one region to another.

tundra
A treeless plain in the Arctic filled with mosses and small shrubs.

TO LEARN MORE

BOOKS

Kallen, Stuart A. *Native Peoples of the Arctic*. Minneapolis: Lerner Publications, 2017.

Regan, Michael. *Bringing Back Our Tundra*. Minneapolis: Abdo Publishing, 2018.

Simpson, Phillip. *Tundra Biomes Around the World*. North Mankato, MN: Capstone Press, 2020.

NOTE TO EDUCATORS

Visit **www.focusreaders.com** to find lesson plans, activities, links, and other resources related to this title.

INDEX

Alberta, 27
Arctic, 5, 7
Arctic foxes, 12

boreal forests, 5–6, 11–12

Canada, 18, 26–28
climate change, 20–21, 29

Déline people, 28

Fortymile herd, 23–24
fragmentation, 19

Idaho, 17

Kaska Dena, 26

mosquitoes, 21

Northwest Territories, 28
Nunamiut people, 14

Porcupine herd, 6

Russia, 18

tundra, 5–6, 11–12, 19

Washington, 17
wolves, 13, 20

Answer Key: 1. Answers will vary; 2. Answers will vary; 3. B; 4. C

Children's 599.658 LON
London, Martha
Saving caribou

05/03/22